SCIENCE EXPERIMENTS
Gravity and Magnetism

Om SPARKS
An Imprint of Om Books International

Om SPARKS An Imprint of Om Books International

Reprinted in 2026

Corporate & Editorial Office
A-12, Sector 64, Noida 201 301
Uttar Pradesh, India
Phone: +91 120 477 4100
Email: editorial@ombooks.com
Website: www.ombooksinternational.com

Sales Office
107, Ansari Road, Darya Ganj
New Delhi 110 002, India
Phone: +91 11 4000 9000
Email: sales@ombooks.com
Website: www.ombooks.com

ISBN: 978-93-52763-08-5

Printed in India

10 9 8 7 6 5 4 3

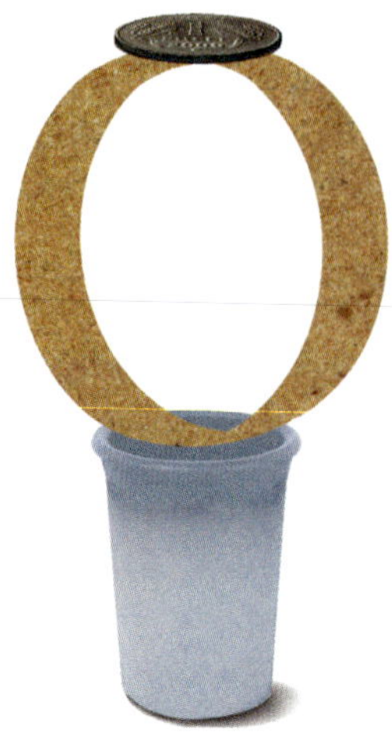

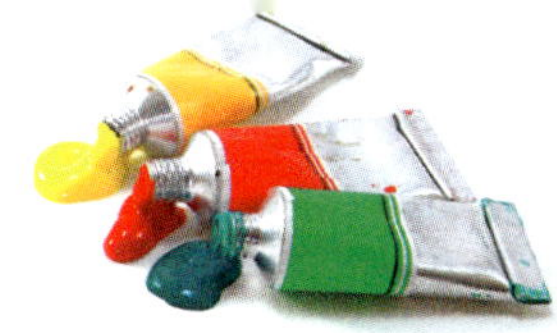

CONTENTS

Duration of Experiment

5 mins

This icon indicates approximately how long the experiment will take, including the set-up time.

Difficulty Level

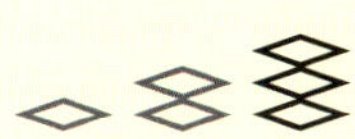

The black bars in this icon indicate level of difficulty in doing this experiment with one icon depicting the least level of difficulty.

Adult Supervision

This icon indicates that parental supervision is required. Always have an adult beside you.

NOTE: All activities and experiments described in this book should be ideally performed with adult supervision. Parents and guardians are advised to oversee the children and assist them with any difficult or potentially harmful tools. Common sense and discretion are advised. The publisher does not assume any responsibility for any injuries or damages arising from any activities.

Tsp - teaspoon Tbsp - tablespoon ml - millilitre l - litre gm - gram 1 cup - 250 millilitre cm - centimetre

Floating Eggs

Duration of Experiment: 10 mins.

Difficulty Level:

What you need:

2 raw eggs

6 tbsp salt

Water

2 glasses

What to do:

Step 1. Add the salt to half a glass of water.

Step 2. Pour tap water till the glass is full. Try not to mix the salt water and plain water.

Step 3. Fill the other glass with tap water.

Step 4. Gently lower the eggs into both the glasses. The egg in the glass with salt will not sink all the way to the bottom. Instead, it will float in the middle of the glass with the salty water.

SCIENCE AROUND US

Salty Seas

It is easier to float in the sea than in a swimming pool because of its salt content. In fact, the salt content in the Dead Sea is so high that it is almost impossible to drown!

What just happened?

The denser salt water remains at the bottom of the glass. It is easy for the egg to float in salt water but not in tap water. The egg sinks through the tap water till it reaches the salt water and floats at that level.

Air Freshener

Duration of Experiment: 2 hours

Difficulty Level:

What you need:

4 packets gelatine

Food colour

Small jars

Hammer

1 tbsp salt

Water

Always take an adult's assistance when you are using the stove.

What to do:

Step 1. Boil one cup of water. Dissolve the gelatine in it.

Step 2. Take it off the heat and add a cup of cold water.

Step 3. Add the fragrance oil, food colour and salt.

Step 4. Pour this mixture into the small jars. Refrigerate for 2 hours.

Step 5. Make holes in the lids of the jars with a hammer and place these around your house to smell the fragrance.

What just happened?

Gelatine is a polymer. To hold its shape, it has a matrix-like structure. The particles of fragrance oil are suspended in this structure. The scent particles are released as the gel evaporates.

Hanging Compass

Duration of Experiment: 15 mins.

Difficulty Level:

What you need:

Pencil

Thread

Large needle

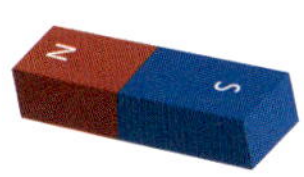

Magnet

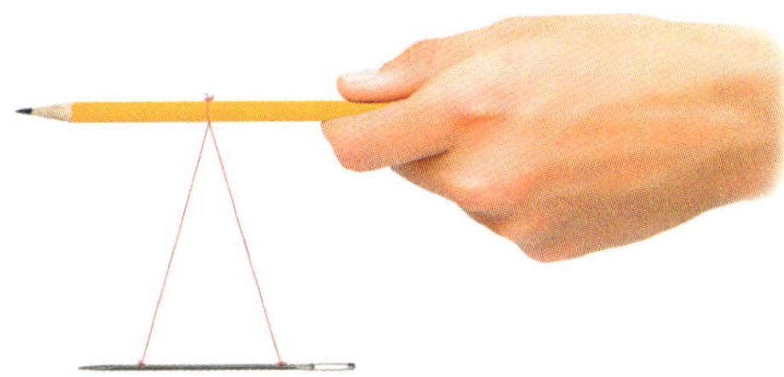

What to do:

Step 1. Stroke the needle with one end of the magnet several times.

Step 2. Tie the needle to the pencil using the thread, as shown in the image.

Step 3. Walk around and watch your needle point to the magnetic North!

What just happened?

By rubbing the needle against the magnet, the needle began to act like a magnet.

Spinning the Bucket

Duration of Experiment: 5 mins.

Difficulty Level:

What you need:

Water

Bucket

What to do:

Step 1. Fill half the bucket with water. Place its handle in the crook of your arm and spin it in a circular motion.

Step 2. If you spin it fast enough, the water will remain in the bucket even when it is upside down.

What just happened?

This experiment uses 'centripetal force'. This force ensures that an object continues travelling in a circular path with a fixed centre.

Wrapped Magnets

Duration of Experiment: 5 mins.

Difficulty Level:

What you need:

Paper clips

Magnet

Cloth

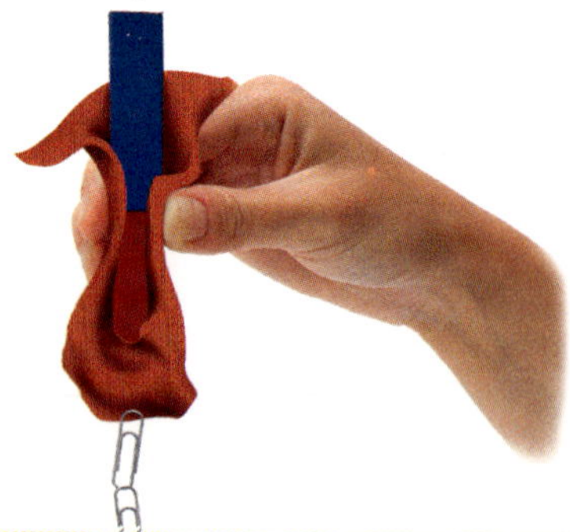

What to do:

Step 1. Wrap the magnet in cloth.

Step 2. Try to attract the paper clips. They will be easily attracted.

Step 3. Wrap the magnet in another layer of cloth and see if the clips are still attracted. How many pieces of cloth does it take for the magnet to stop being effective?

What just happened?

Magnets retain their magnetism even through material like cloth, as long as it is not too thick. Thicker materials prevent the magnet from exerting its force.

Vinegar Mix

Duration of Experiment: 5 mins.

Difficulty Level:

What you need:

3 tsp mustard

Salt

3 tsp cooking vinegar

3 tsp oil

3 glasses

What to do:

Step 1. Pour vinegar, oil, mustard and a pinch of salt in one glass.

Step 2. In the second, mix vinegar, mustard and a pinch of salt. Add oil while mixing.

Step 3. In the third, mix oil, vinegar and a pinch of salt.

Step 4. Try to observe the best mix.

What just happened?

In the first glass, the oil forms small bubbles in the vinegar. In the second and third glasses, the oil and vinegar don't mix. Oil and vinegar don't mix unless the mustard helps bind it.

Musical Buttons

Duration of Experiment: 10 mins.

Difficulty Level:

What you need:

Button

String

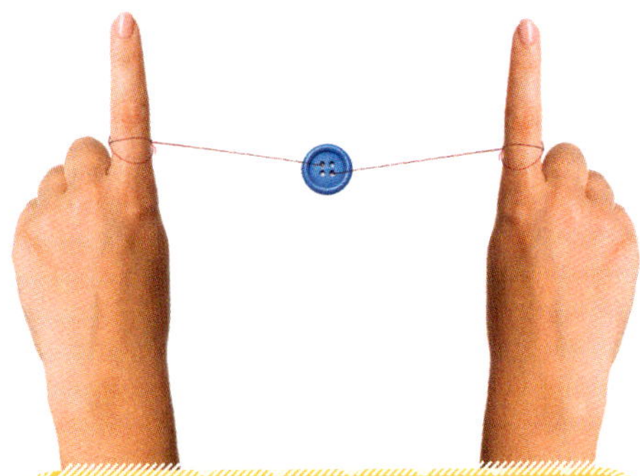

What to do:

Step 1. Thread the string through one button hole and bring it back up through the other. Tie the ends to form a loop.

Step 2. Bring the button to the centre of the string and stretch it out between both your index fingers.

Step 3. Wind up the string. Pull the string tight and loosen it.

What just happened?

The spinning button and the string are vibrating the air around them, making the singing sound.

The Merry-go-round

Duration of Experiment: 30 mins.

Difficulty Level:

What you need:

Needle

3 corks

Bottle

Aluminium plate

Knife

4 Forks

What to do:

Step 1. Slice 2 corks down the middle and stick a fork, at slightly less that 90° into each slice.
Step 2. Balance these on the edge of the plate.
Step 3. Push a cork into the bottle and a needle into the cork.
Step 4. Carefully balance the plate on the needle. With a little trial and error, you can make it stay.

What just happened?

When you balance the objects on an irregularly shaped object perfectly, you have found the centre of gravity for those objects.

Hot Stuff

Duration of Experiment: 15 mins.

Difficulty Level:

What you need:

Small glass bottle

Red food colour

Glass bowl

String

Hot water

Cold water

What to do:

Step 1. Pour cold water in the glass bowl.
Step 2. Fill the small glass bottle with hot water. Add red food colour.
Step 3. Slowly suspend this small bottle into the glass bowl with the help of the string.
Step 4. You will notice the red water rising from the bottle.

What just happened?

Hot water rises because the density of hot water is less than that of cold water, making it lighter. Hence, the red coloured hot water rises and settles above the cold water.

Light it Up

Duration of Experiment: 30 mins.

Difficulty Level:

What you need:

1 battery (4.5 volt) | 1 paper clip | Thumbtack | Light bulb | Tape | Cardboard | 3 wire lengths

What to do:

Step 1. Remove half an inch of the insulation from each end of all three wires.

Step 2. Tape one end of one of the wires to the positive terminal of the battery. Tape the other end to a paper clip.

Step 3. Use a thumbtack to secure the end of that paper clip to the cardboard.

Step 4. Wrap one end of another wire to the side of the light bulb and the other end to the negative terminal of the battery.

Step 5. Wrap one end of the third wire around the negative terminal of the battery and place the other end near the paper clip.

Step 6. Touch the paper clip to the end of the third wire to see the bulb light up. You can now use the paper clip as a switch to turn the bulb on and off.

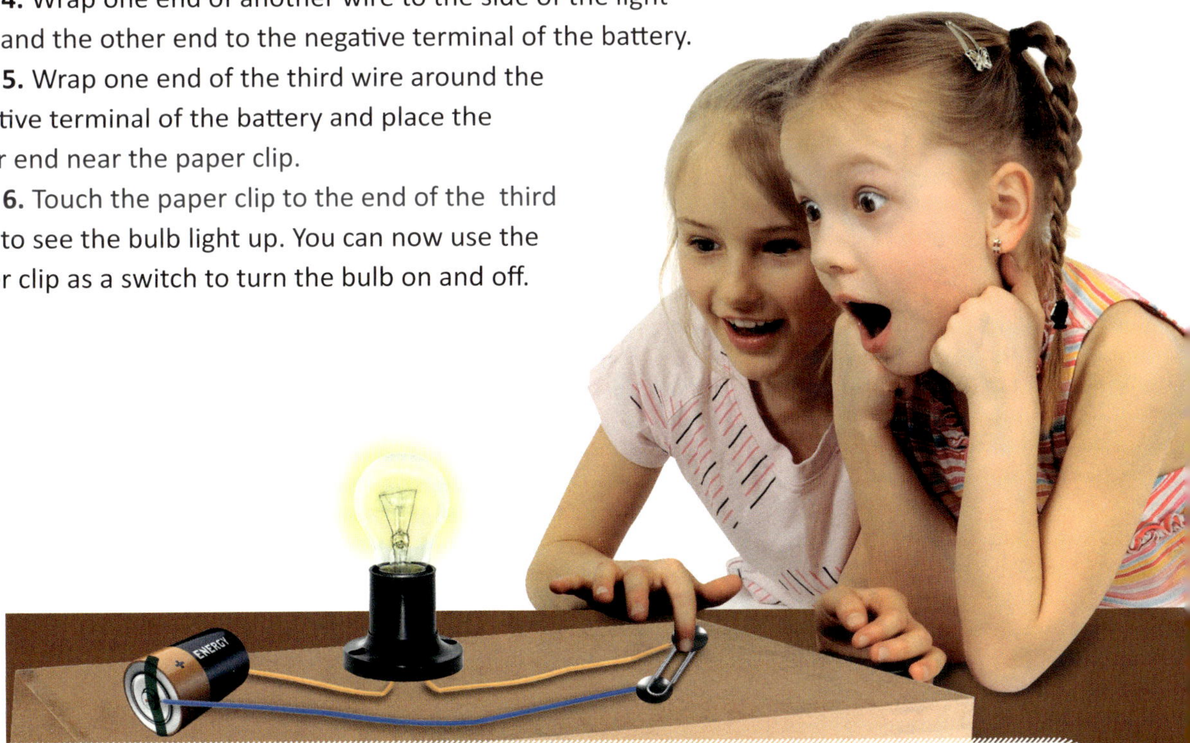

What just happened?

All the materials used in the circuit are good conductors of electricity as they allow electrons to flow freely through them. This free flow of electrons provides energy, which lights up the bulb.

All Uphill

Duration of Experiment: 5 mins.

Difficulty Level:

What you need:

Paper clip

Magnet

Cardboard

What to do:

Step 1. Fold the bottom of the cardboard and make it stand at a slight inclination.

Step 2. Place the paper clip at the bottom.

Step 3. Place your magnet behind the cardboard and guide the paper clip uphill.

What just happened?

Magnetic force works even through non-magnetic materials like cardboard.

Oily Ice

Duration of Experiment: 30 mins.

Difficulty Level:

What you need:

Food colour

Vegetable oil

Baby oil

Glass

Ice cubes

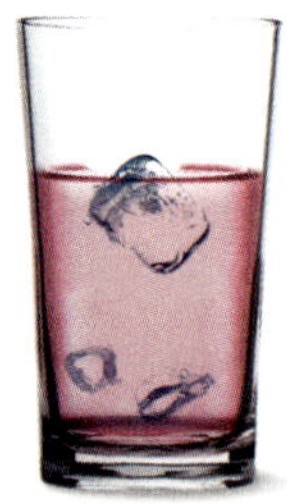

What to do:

Step 1. Add two drops of food colour to an empty glass.

Step 2. Fill half the glass with vegetable oil and the rest with baby oil.

Step 3. Drop an ice cube in it. It settles in the middle of the glass.

Step 4. After 10 minutes, you will be able to see a drop begin to melt away from the ice cube.

Step 5. It settles down in slow motion and mixes with the food colour.

What just happened?

The baby oil settles on top of the vegetable oil because it is low in density. The ice settles in the middle because it is more dense than the baby oil, but lesser than vegetable oil. However, water is more dense than ice and sinks to the bottom.

Make a Compass

Duration of Experiment: 15 mins.

Difficulty Level:

What you need:

Pencil

Clay

Horseshoe magnet

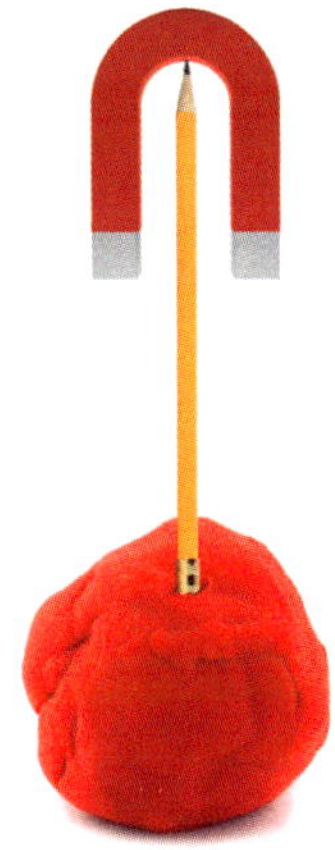

What to do:

Step 1. Mould the clay to form a mound.

Step 2. Stick the pencil into this mound with the pointed end on top.

Step 3. Balance the magnet on the pointed end and let it settle in the north-south direction.

What just happened?

One end of every magnet is called the North Pole and the other, the South. The North Pole of the magnet is always attracted to the South Pole of Earth.

Magnetic Pick-up

Duration of Experiment: 15 mins.

Difficulty Level:

What you need:

Paper

Paper clip

Thread

Tape

Scissors

Magnet

Table

What to do:

Step 1. Cut a 3 inch paper in the shape of a kite. Attach a paper clip to one corner.

Step 2. Tape a piece of thread on the opposite corner.

Step 3. Tape the other end of the thread to a table.

Step 4. Hold the magnet near the paper clip. Watch the kite 'fly'.

What just happened?

Magnets have a magnetic field, an invisible force that attracts some kinds of metal. Metal objects within the magnetic field do not have to touch a magnet to be pulled by magnetism.

Navigation

Duration of Experiment: 15 mins.

Difficulty Level:

What you need:

Needle Cork Pen Magnet Bowl Compass Knife Water

What to do:

Step 1. Stroke the needle with the magnet in the same direction multiple times. This will magnetise the needle.

Step 2. Cut a 1.5 cm slice from the cork.

Step 3. Push the needle in through the diameter.

Step 4. Fill the bowl half with water.

Step 5. Place the cork in the water. It will spin till one end of the needle points to the North.

Step 6. Use the compass to figure out which end is pointing to the North.

Step 7. Mark the North, South, East and West respectively.

SCIENCE AROUND US

Discovery of Magnets

Legend says that a shepherd named Magnes discovered natural magnets 3,000 years ago. He was herding his sheep when the nails on his shoes and the metal tip of his staff suddenly got stuck to the rock he was standing on.

What just happened?

Rubbing the needle with the magnet 'magnetised' it or gave it the properties of a magnet. We know that we can use magnets to find out which direction the North lies in.

Jumping Nails

Duration of Experiment: 5 mins.

Difficulty Level:

What you need:

5 nails

Magnet

Bottle

Oil

What to do:

Step 1. Fill half the bottle with oil and drop the nails in it.

Step 2. Hold a magnet outside the bottle.

Step 3. Watch the nails jump through the oil towards the magnet.

What just happened?

Magnets retain their properties even through oil.

Snake Charmer

Duration of Experiment: 5 mins.

Difficulty Level:

What you need:

Bunch of keys

Thin rope

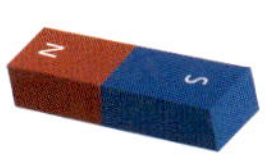

Magnet

Tape

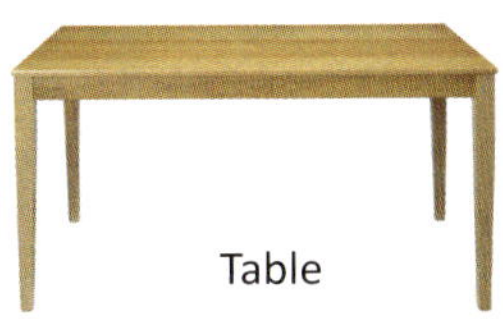

Table

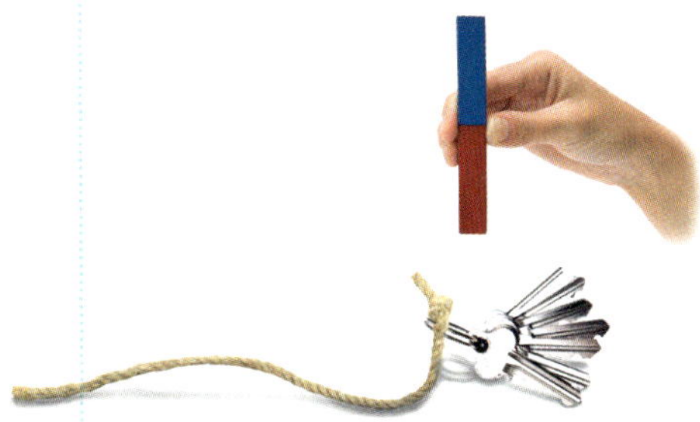

What to do:

Step 1. Tie one end of the thin rope to the keys.

Step 2. Tape the other end to the table.

Step 3. Use the magnet to guide the keys without touching them.

What just happened?

There is an area around the magnet called the magnetic field, where the magnet's force can be felt. The keys need not be in contact with the magnet to be attracted to it.

Lines of Force

Duration of Experiment: 5 mins.

Difficulty Level:

What you need:

Horseshoe magnet

Iron fillings

Papers

Clay slab

What to do:

Step 1. Mould the clay around the magnet such that the magnet can stand with its poles facing upward without support.

Step 2. Place the paper horizontally across the poles.

Step 3. Sprinkle iron fillings on the paper.

Step 4. Watch the fillings arrange themselves into a fascinating pattern.

What just happened?

The iron fillings arrange themselves along the 'lines of force'. These are imaginary lines that show the direction of the magnet's attraction.

Car Race

Duration of Experiment: 15 mins.

Difficulty Level:

What you need:

Rubber bands

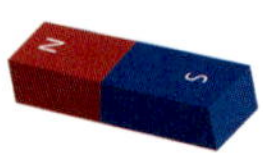

Magnets

Toy cars

What to do:

Step 1. Use the rubber bands to attach one bar magnet under each of the cars.

Step 2. The North Pole should be facing the back of one car and the front of the other.

Step 3. Put the two cars one behind the other and see if you can 'push' the car in front without touching it.

What just happened?

Similar poles of magnets repel each other.

Make a Magnet Painting

Duration of Experiment: 15 mins.

Difficulty Level:

What you need:

Paper clips

Magnet

Paper plate

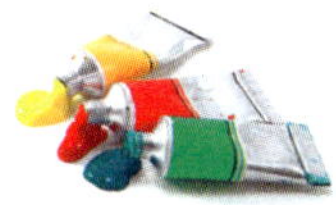

Poster paints

What to do:

Step 1. Squirt the coloured paints onto the plate.

Step 2. Place the paper clips on the plate.

Step 3. Hold the magnet under the plate and guide the clip through the paint.

What just happened?

The paper clip is attracted to the magnet. When you move the magnet, you move the paper clip along with it, creating a metal 'paintbrush'.

Tight-rope Dancing Bottle

Duration of Experiment: 15 mins.

Difficulty Level:

What you need:

String

Glass bottle

Umbrella

Chalk

What to do:

Step 1. Tie the string loosely between any two posts and spread the chalk on it by rubbing against it.

Step 2. Push the handle of the umbrella into the neck of the bottle.

Step 3. Lay the bottle on the string and watch it balance.

What just happened?

The weight of the umbrella and the bottle balance each other out.

Make an Electromagnet

Duration of Experiment: 15 mins.

Difficulty Level:

What you need:

Iron nail

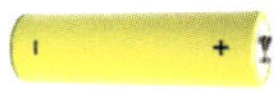
Battery

Thin coated copper wire

Paper clips

Scissors

What to do:

Step 1. Leave about 8 inches of wire loose at one end, and wrap the rest of it around the nail. Try not to let the wire overlap.

Step 2. Cut the wire so that there are another 8 inches loose at the other end too.

Step 3. Now, remove an inch of the plastic coating from both ends of the wire. Attach one wire to one end of a battery and the other wire to the other end.

Step 4. Put the point of the nail near a few paper clips; it should pick them up!

What just happened?

The electricity flowing through the wire arranges the molecules in the nail so that they are attracted to certain metals.

Dancing Raisins

Duration of Experiment: 5 mins.

Difficulty Level:

What you need:

Glass jar

Raisins

¾ cup cooking vinegar

2 tsp baking soda

Water

What to do:

Step 1. Fill $^{3}/_{4}$th of the jar with water.

Step 2. Add vinegar and baking soda to the jar.

Step 3. Drop 10 raisins into the jar.

Step 4. Soon, they start 'dancing' in the jar, sinking and rising.

What just happened?

The vinegar and baking soda reacted to form carbon dioxide, which stuck to surface of the raisins and carried them up. Once the gas reached the surface, it escaped into the air and the raisins sank again.

Air Surfing

Duration of Experiment: 10 mins.

Difficulty Level:

What you need:

5 balloons

Table

What to do:

Step 1. Blow the balloons.

Step 2. Turn a small table over on them.

Step 3. You will find that you can stand comfortably on the table, even with another person!

What just happened?

When you stand on the table, your weight is distributed among all the balloons, which is why they don't burst.

Balancing Act

Duration of Experiment: 30 mins.

Difficulty Level:

What you need:

2 knives

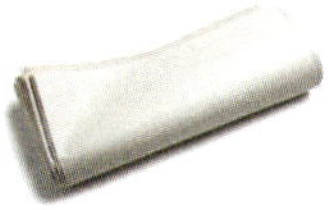
Paper

Cup

What to do:

Step 1. Roll the paper tightly.

Step 2. Use the paper to fix two knives within the handle of the cup as shown.

Step 3. Balance the cup on your fingertip.

What just happened? The knives help distribute the weight of the cup evenly, which makes it possible to balance.

Find the North

Duration of Experiment: 15 mins.

Difficulty Level:

What you need:

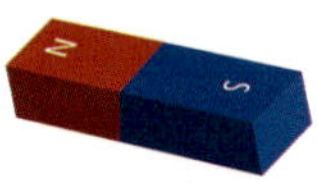
Magnet

String

Compass

What to do:

Step 1. Tie a loop around the centre of the magnet with the string. Suspend the magnet by holding the string.

Step 2. Wait till it settles in one position.

Step 3. With the help of a compass, see which end of the magnet is pointing North. Mark it 'S'. This is the South Pole of the magnet, which will always be attracted to the North Pole of Earth.

Step 4. Simply suspend the magnet anytime; 'S' will always point to the North.

What just happened?

One end of every magnet is called the North Pole and the other, the South. The North Pole of the magnet is always attracted to the South Pole of the Earth and vice versa.

Extract Iron from Your Cereal

Duration of Experiment: 15 mins.

Difficulty Level:

What you need:

Cornflakes

Bowl and spoon

Magnet

White paper

What to do:

Step 1. Crush your cereal in your bowl with the help of your spoon until it becomes a powder.

Step 2. Pour this powder onto the white paper.

Step 3. Run your magnet over it.

Step 4. You should be able to pick up tiny black particles.

Step 5. These are actually little bits of iron in your cereal!

What just happened?

Iron is a mineral that is necessary for our survival in small proportions. It is also magnetic and so gets picked up by the magnet.

Magnetic Race Track

Duration of Experiment: 15 mins.

Difficulty Level:

What you need:

Pencil

Toy car

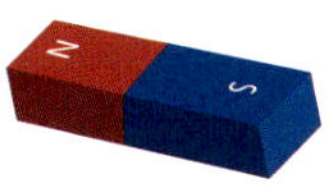

Magnets

Cardboard

What to do:

Step 1. Draw a racetrack on the cardboard.

Step 2. Place the magnet under the car.

Step 3. Hold another magnet under the cardboard and use it to guide the car through the racetrack.

What just happened?

Magnetic attraction works even through the cardboard. That is why you were able to guide the car with the magnet.

The Egg That Won't Roll

Duration of Experiment: 15 mins.

Difficulty Level:

What you need:

Pin

Iron fillings

Raw egg

Glue

What to do:

Step 1. Poke a hole at one end of the egg using the pin.

Step 2. Carefully, make a slightly bigger hole at the other end of the egg by cracking it slightly.

Step 3. Blow through the pinhole so that the eggshell is hollow.

Step 4. Let it dry for a while.

Step 5. Glue the pinhole shut.

Step 6. Through the other end, pour the iron filings in.

Step 7. Pour glue over the filings to hold them in place. Let it dry.

Step 8. Now glue the other hole shut too.

Step 9. Once it's dry, no matter what position you put it in, it will always return to its original upright position!

What just happened?

The iron filings make the egg heavier to where it is glued. This makes the egg return to that position.

Motion Ocean

Duration of Experiment: 15 mins.

Difficulty Level:

What you need:

Glitter

Plastic floating toys

10 drops food colour

Jar

Baby oil

Water

What to do:

Step 1. Fill half the jar with water. Add the food colouring and glitter.

Step 2. Add baby oil till ¾th of the jar is full.

Step 3. Place a floating toy on top of the oil. Screw the lid on.

Step 4. Shake the jar gently to set your ocean in motion.

What just happened?

Water is denser than oil. The two liquids never mix. When the water moves, it pushes the oil around, making shapes like waves.

Magnetic Circuit

Duration of Experiment: 10 mins.

Difficulty Level:

What you need:

Paper cutter

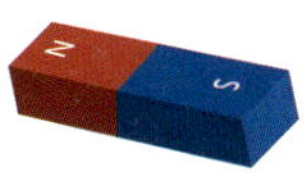

Bar magnet

Compass

4 ft copper wire

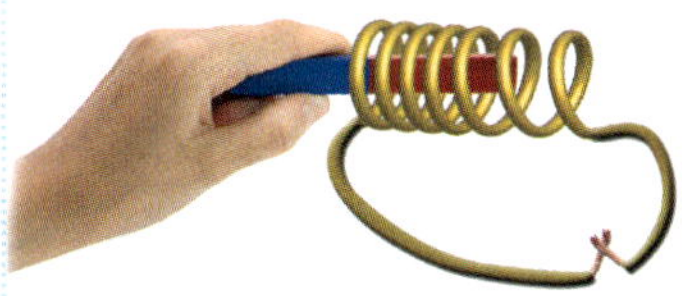

What to do:

Step 1. Scrape the insulation from both ends of the wire.

Step 2. Leave a foot on each end and coil the middle around four fingers.

Step 3. Join both ends of the wire.

Step 4. Move the magnet in and out of the coil.

Step 5. Hold the compass near the coil. The needle of the compass jiggles.

What just happened?

Placing a wire across a moving magnetic field causes a current to flow through the wire. Such a path is called a circuit.

Fun With Magnets

Duration of Experiment: 10 mins.

Difficulty Level:

What you need:

Paper clips

Chalk powder

Salt

Magnet

Plastic bottle

What to do:

Step 1. Add an equal amount of coloured chalk powder and salt in a bottle.

Step 2. Add the paper clips and shake the bottle.

Step 3. Hold up a magnet to the bottle and watch the paper clips jump up towards the magnet!

What just happened?

The metallic objects were attracted to the magnet, but the salt and chalk was not. Only metals are attracted to magnets. That is why the clips 'jumped' out through the salt and chalk.

From Outer Space

Duration of Experiment: 30 mins.

Difficulty Level:

What you need:

Magnet

Jar

Paper

What to do:

Step 1. Collect dust from different locations of your house in a jar.

Step 2. Pour this dust onto the paper.

Step 3. Hold the magnet beneath the paper and tap the paper so that all the dust is on top on the magnet.

Step 4. Tilt the paper while keeping the magnet stationary. Some dust particles will be attracted to the magnet.

What just happened?

Tons of space dust and debris blast Earth everyday. Most perish in the atmosphere, but the few that survive are magnetic in nature. This is exactly what got attracted to the magnet.

Read a Credit Card

Duration of Experiment: 10 mins.

Difficulty Level:

What you need:

Rusted nail

Old credit card

Sandpaper

What to do:

Step 1. Scrape some rust off the nail with the sandpaper.

Step 2. Pour this rust over the magnetic strip of the card.

Step 3. Tap off the excess rust. Look at the strip under uniform white light.

Step 4. Some black particles from the rust will be stuck to the magnetic strip.

What just happened?

The magnetic strip on a credit card stores its information magnetically, so areas of the strip will be magnetised. The black particles from rust contain iron, which is attracted to the magnetic strip.

Hot Magnets

Duration of Experiment: 25 mins.

Difficulty Level:

What you need:

Pliers

Needle

Compass

Magnet

Be careful while heating the magnet. Always make sure that there is an adult around to supervise when you use fire.

What to do:

Step 1. Use the bar magnet to magnetise the needle. This can be done by stroking the needle with the bar magnet in the same direction several times.

Step 2. Hold it near a compass. You will see the compass needle move.

Step 3. Hold the needle over a hot stove with the help of the pliers.

Step 4. Hold it there for 20 minutes.

Step 5. Now, take it off the flame and hold it near the compass. It will have no effect.

SCIENCE AROUND US

Why does a magnet lose its magnetism?

You can imagine a magnet as a bunch of tiny particles pointing in the same direction. When you heat a magnet, the little magnetic parts can get jostled and unaligned, destroying its magnetism.

What just happened?

The first time you held the needle near the compass, the needle was magnetised and made the compass needle move. After heating, it lost its magnetic properties and had no effect on the compass.

Magnetic Poles

Duration of Experiment: 5 mins.

Difficulty Level:

What you need:

Iron fillings

Magnet

White paper

What to do:

Step 1. Place the bar magnet on the piece of paper.

Step 2. Sprinkle the iron fillings around the magnet.

Step 3. Try to see where there are maximum iron fillings.

Step 4. Most of the fillings are attracted to the two ends or poles.

What just happened?

Each magnet has a magnetic field or an area where its force is felt. This field or area of influence is strongest at the two poles.

Thumps Up!

Duration of Experiment: 5 mins.

Difficulty Level:

What you need:

Long ruler

What to do:

Step 1. Put your hands out in front of you and balance the ruler on them.

Step 2. Now slowly move your hands closer to each other.

Step 3. The point at which both your hands meet is the centre of gravity. The ruler will balance at this point.

What just happened?

Because of friction between the hand and the ruler, it is possible to determine the exact centre.

Parachute

Duration of Experiment: 30 mins.

Difficulty Level:

What you need:

Scissors

String

Small toy

Plastic bag

What to do:

Step 1. Cut out an octagon from the plastic bag.

Step 2. Cut a small hole near the edge of each side.

Step 3. Attach eight pieces of equal-sized string to each of the holes.

Step 4. Tie the pieces of string to the toy and drop your parachute slowly.

What just happened?

When you release the parachute, the weight pulls down on the strings and opens up a large surface area of material that uses air resistance to slow it down.

Levitation

Duration of Experiment: 15 mins.

Difficulty Level:

What you need:

Tape

Round magnets

Scissors

Card paper

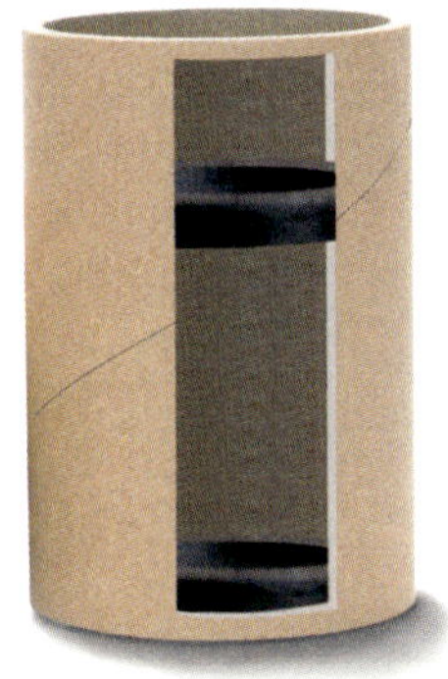

What to do:

Step 1. Turn the card paper into the shape of a tube and tape it. The diameter should be slightly larger than the round magnets.

Step 2. Slit the side of the tube so that you can look inside clearly.

Step 3. Put both magnets in the tube with both north poles facing each other.

Step 4. Look at the levitating magnet through the slit!

What just happened?

Similar poles of magnets repel each other. The repulsion caused one magnet to levitate.

Race to the Bottom

Duration of Experiment: 5 mins.

Difficulty Level:

What you need:

2 objects of different weights
(like a bottle cap and a coin)

Chair

What to do:

Step 1. Stand on a chair.

Step 2. Drop both objects at the same time.

Step 3. Try to see which one reaches the ground first.

What just happened?

Both reached at the same time! Though you may think that the heavier one would reach first, gravity exerts the same force on all objects.

Floating Magnets

Duration of Experiment: 15 mins.

Difficulty Level:

What you need:

7 corks

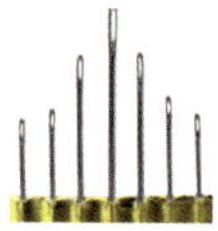
7 needles

Magnet

Container

Water

What to do:

Step 1. Fill the container with water.

Step 2. Place all the needles together in a row. Make sure they are all facing the same direction.

Step 3. Magnetise them by stroking them with one pole of a magnet.

Step 4. Push the needles through the top of the corks. Let only a little bit of the eye stick out.

Step 5. Place these in the container.

Step 6. Hold a magnet over it and watch the needles form patterns.

What just happened?

All the magnetised needles have North and South Poles. They are therefore either attracted to or repelled by the magnet. However, because they are magnetised, the needles also repel each other.

Roly-Poly Doll

Duration of Experiment: 30 mins.

Difficulty Level:

What you need:

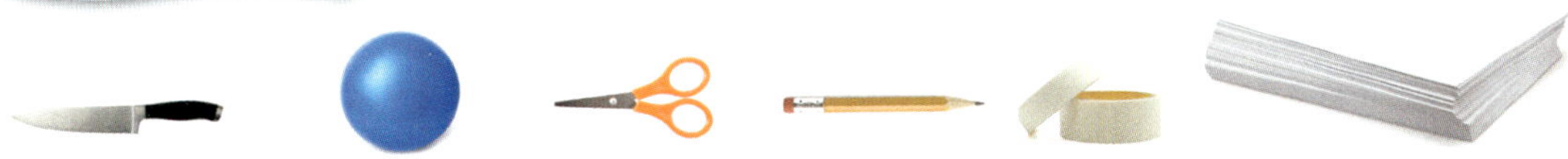

Knife | Bouncy ball | Scissors | Pencil | Scotch tape | Paper

What to do:

Step 1. Cut the bouncy ball in half.
Step 2. Place it on a table with the flat side facing up.
Step 3. Draw a 10 cm × 10 cm square on the paper. Cut it out.
Step 4. Roll the square paper to form a tube. Tape it to the cut ball. The round part of the ball should stick out from the bottom.
Step 5. Keep cutting small sections off the top of the paper tube till the tube returns to an upright position when it's knocked over.
Step 6. Draw a clown on the paper tube and play with your toy.

What just happened?

The centre of gravity for an object is the point at which the object's average weight is located. If you support the centre of gravity, you support the entire object. When you cut this ball in half there is a shift in the centre of gravity and the object loses its support, which causes the toy to return to an upright position after being knocked over.

The High Dive

Duration of Experiment: 15 mins.

Difficulty Level:

What you need:

Coin

Card paper

Scissors

Pencil

Water

Film canister

What to do:

Step 1. Cut the card paper into a long strip to form a hoop.

Step 2. Put a little water in the canister. Place the hoop on the film canister and balance the coin on it.

Step 3. In one swift motion, use the pencil to fling the hoop off to the side.

Step 4. The penny will fall straight into the canister with a splash.

What just happened?

When you swiftly got the hoop out of the way, the penny was left suspended over the film canister with nothing to hold it up. Gravity caused it to fall straight into the canister.

Swing King

Duration of Experiment: 5 mins.

Difficulty Level:

What you need:

Watch

Swing

What to do:

Step 1. Hold the seat of the swing and move behind by pushing back 2–3 steps using your feet. Form a mark in the ground in front of your feet.

Step 2. Let go of the ground and let the swing more forward. Count the number of times the swing goes back and forth in 10 seconds using the watch. Your friend can help you keep time.

Step 3. Now, ask your friend to sit on the swing.

Step 4. Pull the swing back to the exact position it was in earlier. (You can do this with the help of the mark you made on the ground earlier.)

Step 5. Let go of the ground like before and count how many times the swing goes back and forth. The swing would have moved the same number of times be both your friend and you.

What just happened?

The swing moves because of gravity's force. This gravitational force does not change if the weight differs.

SCIENCE AROUND US

Pendulums

Pendulums are designed in away such that once they start moving, they keep moving. The basic force that makes a pendulum move is gravity. However, the pendulums in clocks use a complex system of gears to keep it ticking.

OTHER TITLES IN THIS SERIES